AN INCOMPLETE HISTORY OF TIME

KEVIN SUN

PARTRIDGE

Copyright © 2021 by Kevin Sun.

ISBN: Hardcover   978-1-5437-6600-4
        Softcover   978-1-5437-6601-1
        eBook   978-1-5437-6602-8

All rights reserved. No part of this book may be used or reproduced by any means, graphic, electronic, or mechanical, including photocopying, recording, taping or by any information storage retrieval system without the written permission of the author except in the case of brief quotations embodied in critical articles and reviews.

Because of the dynamic nature of the Internet, any web addresses or links contained in this book may have changed since publication and may no longer be valid. The views expressed in this work are solely those of the author and do not necessarily reflect the views of the publisher, and the publisher hereby disclaims any responsibility for them.

Print information available on the last page.

To order additional copies of this book, contact
Toll Free +65 3165 7531 (Singapore)
Toll Free +60 3 3099 4412 (Malaysia)
orders.singapore@partridgepublishing.com

www.partridgepublishing.com/singapore

CONTENTS

An Introduction to Time..........................ix

What is the Past?..................................1
What is the Present?.............................10
What is the Future?..............................15
What is Entropy?.................................25
Is time Real or just an Illusion?...............29
What is the Direction of time?................40
The Speed of Time..............................45
Time Travel..62
Conspiracy Theories Surrounding
    Time Travel.....................................82

To those that I love until
the end of time . . .

AN INTRODUCTION TO TIME

First things first, this is not the same as the book *A Brief History of Time* written by Stephen Hawking. This book is "an incomplete history of time," and it is written by a thirteen-year-old boy—me. This book is not really about history, but it is about the only reason why history even exists—time. If you are trying to use this book as a tool to help you with your science exam, I suggest you put it down right now. This book probably has many flaws and mistakes, and please correct me if I am wrong, however, I am only recording my own understanding of what time is, what it may be and what it could be. There is one thing that I am quite confident about: Even if you throw this book away, you will not find any information that can truly define and explain what time is, either online or offline, because throughout human history, nobody, including the most brilliant minds that have ever existed, can answer this simple yet complicated question.

What is time? When I throw this question at you, you may say time is the past, the present, and the future combined, and that is correct, but what is it really? You may think you already know what time is: sixty seconds is a minute, sixty minutes is an hour, twenty-four hours is a day, and so on. However, that is not time itself, simply units we have come up with to measure it. Time has been an old friend to us humans for thousands, if not millions, of years, and generations before us have made enormous progress on understanding time physically. Albert Einstein has discovered the theory of relativity that suggests the world as we know it is a fabric/grid of time and space. Yet most people still cannot define it in a noncontradictory way because we haven't understood it philosophically.

In case you haven't noticed, the picture below the title on the cover page (the arrow with a dotted line in front of it) is the image that appears in my head when I think of the arrow of time. The arrow of time is quite similar to a timeline, and it can almost represent and resemble the functions and the features of time completely. Let me explain: The black line on the left to the arrow represents the past, which has started since the big bang (we'll elaborate on that in the next chapter), the arrow in the middle represents the never-stopping present, and the dotted line on the right of the arrow (which I added myself) is the future, which hasn't happened yet. Does the arrow of time always point to the future?

Presumably not because, really, our future is just our future past and our past future. Since nobody in the past or the present is really sure how time operates, scenarios and predictions like this actually stand.

Time is so hard for us to define and understand because time is not the same as an apple. It doesn't have a physical form (at least for now). All the images that appear in your mind when you think of time are only representations of it, such as a clock, a calendar, or a timeline. We cannot see, smell, hear, taste, nor touch time, and most of the time we fail to notice it. Time may not seem as real as this book in your hand, but it never fails to deliver its effects on us, and that is why we age every second we live. Because of time, the sky lights up each morning, it grows dark each evening, and everything operates the way we experience it. Maybe it's not such a mystery after all.

In the following chapters, you and I, together, are going to unveil the secrets of time, find answers to the questions *What exactly is the past, the present, and the future? Is time real or just an illusion? What is the theory of relativity? How to solve the unsolvable paradoxes of time travel?* and a lot more. Hopefully, we will also come up with new questions along the way.

WHAT IS THE PAST?

Before we dig into the past, we need to know what the word *past* means; the word *past* is the noun form of *pass*, and this makes perfect sense in our case because time is always passing, second by second, and we cannot stop it. The past is the period that has passed, and since time is always pushing itself forward, nonstop, the time the past covers is always increasing as well (if that makes sense to you). To make it clearer, the past is a series of events that have happened, occurred before the present and the future. For example, you drank a cup of coffee yesterday for breakfast is something that happened in the past because the coffee you drank is no more; it has been digested, its nutrients absorbed by your body, and now, once a cup of freshly-ground coffee is nothing more than a few hundred millimeters of urine. Even if you order the same coffee from the same cafe again today, your newly-bought coffee may look similar, taste similar, but it isn't the same as the coffee you drank yesterday because remember, you drank it, didn't you? Therefore, we can come to the conclusion that what has happened has happened, we cannot (yet) change the past (no matter how desperate we are), and that specific cup of coffee is lost in the past forever. Piece of advice here, most people are still daydreaming about what would happen if they could change the past so that they would be living in a better present when they should really be working on creating the future to be a better place or rather a better time.

I personally believe there are four different *pasts*; those are your past, my past, our past, and the past. Your past is the series of things that you experienced in the past from your perspective, which means it is mixed with your feelings, emotions, observance, and understanding of those events. My past is the series of things that I experienced in the past from my perspective, which means it is mixed with my feelings, emotions, observance, and understanding of the events. My past and your past are ultimately different, and by different, I mean by very different. We all have different childhoods, families, friends; we all came from different places, experienced different things, met different challenges, achieved different accomplishments; and after all, each of us is very different people from the other with very individual minds. Our past is the series of events that happened which we all experienced at the same time. For example, we are neighbors living in the same suburb, and one day an ice cream truck arrived on the street, and we saw it at the exact same time, and that makes the memory of the ice cream truck a part of our past. Our past is something that happened to you and me at the same time, and our past only exists when there are two or more witnesses. THE (pronounced "zi") PAST is the series of events that has happened across the universe, and THE PAST is the most "accurate" past of all four *pasts* and the most precise account of past events. Look at THE PAST like this: It is an unbiased, complete, and accurate record of THE PAST, and the other three *pasts* are more self-centered and probably not so accurate because of the influence of the people experiencing them. However, THE PAST only exists in our thoughts because when we experience or simply view things, we cannot stop ourselves from being overwhelmed by feelings and cannot help but to add a bit of ourselves to what was. Maybe only God himself can truly see THE

PAST if there is a God. So? Which one of the four *pasts* is real? Or could they all be real?

I believe we are all familiar with the fact that the past determines the present and the present determines the future. It all makes perfect sense, right? However, there is a plot hole; it did not indicate what determines the past. You may think, of course, the past determines the past, but you may also think, well, there must have been something before the past. To fill in that plot hole we have to jump on a time machine and go back 13.7 billion years ago, back to the birth of our universe, the big bang. It is the common belief that the world as we know it was only created after the big bang banged, and it has been expanding far and wide ever since. Basic elements came into existence not long after the big bang, and little by little, our universe was shaped. I think it is fair to say that without the big bang, the book in your hand wouldn't exist, the clothes you're wearing wouldn't exist, and you wouldn't exist. Back to the topic of time, more and more scientists (including Stephen Hawking), physicists, and cosmologists are supporting the theory that time itself was only created after the big bang, along with countless other things. If this theory is true, then it answers our question, and there was no time before the big bang, no past before the past, and this means the past only started shortly after the big bang just like time itself, even though the period of the past is increasing at a very fast rate every millisecond as you read through this text, it is still finite, and the big bang marks its start. But what was there before the big bang, you may ask. Well, scientifically, there was a dimensionless singularity, but I would like to think of *whatever it was* as a place without time, light, motion, change, life, death, love; a place without anything but nothing.

It is undeniable that the past is always a major and important part of our lives because, after all, our past directly affects our present and indirectly influences our future. Hear me out, imagine you are an innocent little lamb living in a farm, and just recently, you wandered into the territory of a wolf. The wolf chased you, and you barely made it back to the farm alive. Now, I ask you, will you go into the turf of that wolf again after having a glimpse of your possible fate? No one in his/her right mind would want to take that plunge again, and we can see how experiences in the past prepare us for challenges in the present and the future. Another example, where did I get the knowledge to write this book? Well, I learned a lot from past encounters, YouTube videos made by YouTubers in the past, and also books written about the past by historians and scientists. I wonder how long it must have taken for the first ape-man to learn that wild beasts are not friends because there was no ape-man before him/her, no guide, no warning; a great price must have been paid. In fact, our past is so important to every one of us because if I take away your memories, recollections of your past, you will not know who you are, where you are, and why you are, that you are not much different compared to a newborn baby. By now, you've probably realized that it is the past that shapes your identity, from your body to your mind.

Sadly, our past discussions about the past have ended quickly, but fear not, the things I wrote, which you read, would always be in your memories, and you can come back to them whenever you want, for they are trapped in the past forever, and they have made a ripple, if not a mark, in your past. At present, you should turn to the next chapter about the present.

WHAT IS THE PRESENT?

Before we dig into the present, we need to know what the word *present* means; the word *present* means currently existing or occurring at this point, but it also has another meaning, a gift, which it is because it allows us to set things right, whether in the past or in the future, and as I would like to quote it: "The present is the easiest and the most simple way to travel to both the future and the past." This quote may not make much sense to you right now, but I hope you will appreciate it by the end of this chapter. The present is unlike the past and the future because the other two are two-dimensional (a line, a period), while the present is one-dimensional (a point, an instant). I believe you have probably played back, paused, and fast-forwarded a YouTube video before, easy, right? In the video's case, the contents you've watched is the past (of the video), the remaining contents is the future (of the video), and the present (of the video) is that little red dot on the progress bar (the video's "timeline"), which moves to the right as time passes.

`0:12 / 2:22`

We can describe the present as a moment where the past meets the future, like corned beef between two loaves of bread; however, because of the present's special position in the arrow of time, it tends to blend into the past as soon as the future mingles into present.

But is the present really just a point in history? Many people argue that the present may be more than just a trice, and they have their persuasive reasons. Both sides of the argument agree upon one thing, and that is the present is not fixed. What do I mean by that? The present you are going through right now is going through change itself, and simply put, it is changing and moving forward on the timeline. Here is where the issue comes in: For change to occur, it requires time, it needs the passage of time, and it always takes a period for something to change, no matter how little the difference is and how short the duration is. For example, you were to bake banana bread, and a freshly-baked banana bread wouldn't magically make its sudden appearance on the dinner table, would it? There are processes and methods you need to follow to turn the ingredients into the actual produce. I wonder how long that would take. I am quite confident to believe that it takes at least longer than an instant. (And if it did only take an instant for you to make banana bread, I suggest you file for the Guinness

World Record right now before someone else does.) Anyway, the point is change is the most basic component of time, and it takes a period for change to happen; the present is always changing, which means the present may not just be a trice but a rather quick, very quick, period. In fact, it is a phase so short that we ignore it completely and don't count it as what it really is. Perhaps our brains just can't process the things we saw, smelled, heard, tasted, and felt as what they really are, a teeny tiny section of the not-so-distant past, like a tenth of millisecond.

I must admit that even myself don't truly understand what exactly the present is, and I don't believe that any human has been able to clearly define the present in the past, and the present. I can't spend so many paragraphs on the present since it is probably one of the most mysterious branches of time;

however, that is not necessarily a bad thing because it allows us to make all kinds of crazy assumptions that may seem extreme, of course, in a hundred years' time. And where does a hundred years' time leave us? The future, of course, which we will be looking at in the next chapter.

WHAT IS THE FUTURE?

Before we dig into the future, we need to know what the word *future* means; the word *future* means the time directly after the present, and it is the last of the three components that make up time. The future is very unpredictable compared to the past and the present because the past and the present are made up of things that have happened and things that are happening, while the future is full of things that haven't happened yet or never will happen. We really don't know what to expect from the future, let alone defining it. To gain some understanding of the future, we have to think about two questions: When did the future start? And is the future infinite? If you have read the previous chapters, you probably know that time only started "ticking" at the instant the big bang banged and the present was created, then the future and the past after a short passage of time. Now, will the future end? It all depends on whether time and the universe ends. The result of our universe's end is unknown; however, many believe that all the matter and space-time in the universe would collapse to what it was before the big bang, a dimensionless singularity where there was no time. So the answer is pretty simple: As long as the universe is untouched in one piece, the future will always be infinite like time. Unlike time itself, we humans, unfortunately, fall victims to this natural law, and our lifetimes are finite (at least for now, who knows what science and biology breakthroughs can be made in the future?). Our life started when we were born, and the countdown started since then, and our life, our own timeline, ends decades later with our death, and by then, we will no longer have a future.

How much does the future mean to us? You may argue, not much, we need to live in the present and focus on what is happening right now. I mean, yeah, you're right, but without the future, you'll only find that the present will end too quickly. A research revealed mind-blowing results of how much time we spend on thinking about each of the three sectors of time. We spend 10 percent of our time thinking about the past, 10 percent thinking about the present, and unsurprisingly, we think about the future all the time. I bet you have thought about or are thinking about what you are gonna do in the future after reading this book, whether planning a date or deciding what to eat for lunch. Because of the effects of the future, we humans like to plan things ahead and prepare for things that may not happen at all, for example, taking an umbrella with you to work on a gloomy morning, and this instinct is in our nature that was developed over thousands of years. This is because we feel uncertain about the future, and we come up with strategies to deal with the outcomes because you never know what is waiting for you in the future, whether winning the lottery or getting infected with Covid-19. You will never find out what will reach you first, tomorrow or a terrible accident. And because of that concern about the result of the future, humans developed two new skills that help us penetrate through the fog surrounding the future. I will come to what those two skills are in a moment, I promise, but I need elaborate on why they existed in the first place. I believe you all wish to have a glimpse of your futures, and well, it's impossible (unless you have some supernatural abilities). That craving and eagerness for knowledge about the future led to the existence of the two abilities. The first skill is prediction, and it is about making assumptions about the future based on facts and knowledge about past or present events, and it greatly increases our preparedness. For example, imagine a man living in a suburb that doesn't lack supermarkets of any kind; we can, therefore, predict that the man doesn't go outside his suburb when he wants to buy groceries, and our prediction has a very high chance of being correct. The second "skill" is divination, like a ritual to foresee, to foretell, to prophesy the future. It is often closely related to religions. While some people doubt the accuracy of this specific subject based on mythological beliefs, palmistry, tea leaves reading, looking through a crystal ball, and observing the constellations, we have to see why many others still pay a witch or wizard-looking person five bucks at a carnival to see what that person has to say about their future after observing an ordinary glass ball for five minutes. Simply put, divination is the efforts humans have made to know something before it happens, to become divine, to become like a god.

Is the future already set? Is the road ahead already paved? It's hard to come up with an answer that everyone will be happy about for questions like these. We will look at these questions from two perspectives: First, the future is set, and second, the future isn't set. If the future is determined, it means we can't change the future, despite our efforts in the present. For example, if you are predetermined to fall off a building at 23 Road tomorrow, then you will 100 percent fall off a building at 23 Road tomorrow. You obviously want to live, so now, after knowing you fate, you do your best to avoid it, to change it; however, all your work will end in vain. One way or another, you will fall off a building at 23 Road tomorrow and die, even if you try to escape, let alone that you can't foretell your future. In this case, our lives are nothing more than progress bars of a recorded video, but doesn't this mean we have already experienced our future in the past because the video was prerecorded? (I know all this is super confusing, but you are not the only one, I don't really get it either.) If the future is not scheduled, it could be far more creative and colorful compared to the determined future. It just means you play a much more vital role in your future, that you could redirect it from its original trajectory. For example, you are the child born to poor farmers, and it is most likely that you will become a farmer as well when you grow up, but no, you studied hard, and eventually, you got your university degree, then you became a rich businessman. You have more power to decide what you want your future to be like; your actions in the past and the present result in consequences, and those consequences are your future. So in this case, you are the boss, you are the king, and you are the director of the movie titled *Your Future*.

Next question, what is the probability of each possibility of the future? Before we try to answer this very "tough" question, we need to know what probabilities and possibilities are; they are similar but not the same. Possibilities are the possible outcomes of an event and how it could result in. For example, when you are reading this book, you could be wearing clothes or not, and they are the only two possibilities regarding your clothing because you can't be wearing clothes and not wearing clothes at the same time. Probabilities are the likelihoods of each of the possible outcomes happening. For example, the two possibilities regarding your diet when you are trying to lose weight would either be eating salad or fast food; however, since you're on a diet, you're more likely to choose the salad over the fries, so your probability of having the salad is relatively higher than that of the burger. Now you know what possibilities and probabilities are, we can go back to answering the question, *What is the probability of each possibility of the future?* There are infinite possibilities of the future. Who knows what could happen tomorrow—maybe World War 3 will begin, maybe King Kong will attack the city, maybe you will become Spiderman, or maybe time travel will no longer be a dream? However, the probabilities of all those events are really unlikely, and I am confident to say that they will (almost) never occur in our lifetimes. According to Murphy's law, "If anything can go wrong, it will." Simply put, if there is a possibility of something happening, no matter how tiny its probability is, over time, it will still happen. For example, the probability of intelligent lives surviving on a specific planet, such as Earth, in this vast universe is quite low, yet here we are. *Homo sapiens* first appeared on Earth 4.6 billion years after its creation. In everyday life, we don't have to rely on possibilities to achieve our goals, but we can increase our probabilities by pursuing our aspirations. At the end of the chapter, I just hope that the person reading this now will have a bright future, and always remember, your actions in the present counts. Perhaps we are the ones writing our own *Hamlet* after all. For now, I'll see you in the next chapter.

WHAT IS ENTROPY?

Let me guess, you've probably never heard of the word *entropy* before in your life, maybe you have in your physics class, and I assume you've forgotten all about it. Entropy is the measure of the amount of disorder or randomness in a system according to the second law of thermodynamics. Before we talk about the second law of thermodynamics, we need to know what the first law of thermodynamics is. It states that any type of energy is always conserved. You cannot create or destroy energy. Instead, one type of energy can only be transformed into another type of energy. For example, when you drive a car, you create kinetic energy (movement), which can be transformed into heat energy and sound energy. The second law of thermodynamics states that as the transformation of different types of energy occurs, more and more useful energy will be "lost," thus resulting in chaos and disorder. The second law of thermodynamics also states that over time, the level of entropy in a conserved space would always increase or stay the same, but it would never decrease. For example, when you are making a cup of hot tea on a cold winter night, have you ever wondered why the tea cools over time? Well, that's entropy's doing. Another example, on a hot summer day, you get an ice cube out from the fridge, and the ice cube is something of low entropy, meaning it is relatively ordered because it's packed in a solid cube shape, but after a while, the ice cube melts into a puddle of water of high entropy and less ordered, and you can make it even more disorganized if you drop drops of that water from the puddle onto random places around the world, and entropy will increase even more. Entropy is the process that our universe is going through to distribute energy equally. If this process continues, then it is predicted that our universe will eventually reach equilibrium, a perfect distribution, in which there is no free energy, no light, no heat, no reaction, and no life, and the universe will return to its original form before the big bang. There are trillions upon trillions of atoms in the world around us, so really, it's more likely to be disordered than ordered. That is why when you want to eat ice cream, the chance of you just happening to find one under your bed is really low that it has never happened before, and in fact, the probabilities are so small that it can be deemed impossible. The larger the system, the less likely it is for entropy to be reversed. That is also why time in the world observed by us only flows in one direction (at least for now). The amount of entropy in our universe today is always higher than that of yesterday because the more changes you make today, the more changes will be made to the changes you made today tomorrow, thus becoming more and more disordered, thus increasing the level of entropy, and once harm is done, it is irreversible. For instance, you get a cut on your hand, it heals in a week from the original hand, then the scab comes off, but your healed hand is never the same. Entropy is related to the arrow of time, which I mentioned before. You can never have the same coffee you had yesterday for breakfast. (Remember that concept?) You may think, all right, now I know what entropy is, what does it have to do with time? Well, that's what we'll discuss in the next chapter.

IS TIME REAL OR JUST AN ILLUSION?

Before we can answer this very controversial question, we need to recap on what time is again. According to Einstein, time is a dimension just like space, and together, they form space-time. Here's an example to help you understand what space-time is: You want to meet up with your friend for a cup of coffee, so you would give them a specific location and a time. If any of the two is missing, your little coffee meeting would be ripped apart. Even though space and time are closely related, there are many differences between the two: Why can move through space in any direction we want (up, down, left, right, diagonal) but in time, only one (forward)? It's really easy to answer this question. You can imagine space as a three-dimensional shape that has its edges stretching out and expanding almost infinitely in every direction into the universe, while time can be interpreted as a two-dimensional line, a timeline, and you can only travel forward or backward in a line, right? However, we are yet to learn how to travel backward or skip forward. We can also think of time as a fourth dimension that we often fail to notice from the original three; this fourth dimension measures the change in the first three. Back to our question, is time real or just an illusion? The issue has been an argument for centuries, and there are always people for and against it. The Greek philosopher Aristotle once proposed that time is relational, that time is the measure of change. For example, if everything in the world we are living in were to suddenly stop, it would feel like time has paused. Another example, do you think you can tell a photo of a dead bear (an instant) from a video of the same motionless dead bear (a period)? Probably not, and it further proves that if there isn't change in the environment around us, we cannot feel the effects of time. Now you may think, okay, time is change, and the one thing that doesn't change is change, so time exists. After all, it's hard to imagine a world without time because we've been introduced to it since the time we were born. Generations before us have studied time, and Sir Isaac Newton made time a physical quantity, and in a way, this proves time is objective (not biased or influenced) and not something subjective (only exists in our minds). Time, as a physical quantity, has appeared in many scientific and mathematical equations that we are familiar with. For example, to calculate the speed of something, we can apply a simple equation, which is distance (kilometer) divided by time (hour). Don't be fooled to think we've just proven that time exists just because it's a vital part of an equation. Newton, the man who introduced time as a physical quantity, is not objective, and all the scientific terms we use these days are nothing more than the value "x" in an algebraic expression. In this case, 1 + 1 doesn't always have to equal 2. It all depends on the meaning you give to the plus sign, and really, the number 1 is as illiterate as time; think of them both as a structure, hollow on the inside.

But what if time is just an illiterate thing that we trap inside clocks and any other measurements of time? To prove that time is an illusion is much easier than proving its existence. Perhaps time is the most common misconception and illusion developed by humanity over millions of years. Since the dawn of time, humans, as a species, have strictly followed a schedule, and time is the fundamental part of the schedule. We wake up in the morning, we then have breakfast, and then we go to work, and so on. Time is a concept created by our minds to track the flow of events in a convenient way. This shows that the very existence of time is subjective, just like our minds, because time is a creation of the mind, and therefore, it can never be objective. Time is a feeling experienced vastly among all life forms, and it seems that we are the ones trapped in time this one-way lane, and all this time, we failed to notice that this world around us is an everlasting prison. We can come to a conclusion that time is life and we are time. The only things that feel the passage of time in this universe are life forms. According to quantum mechanics time is somewhat similar to the concept of Schrodinger's cat, where the particles are in an uncertain state when there is no one to observe it. The same thing may have happened with time, and in a way, time was only created when life first existed. Furthermore, our understanding of time isn't so convincing or complete. NASA organized an experiment on circadian rhythms, where they sent Stefania Follini, an Italian woman, into an underground cave to live, and the effects on her body and mind were recorded. In the cave, there was an abundant supply of food and water and lots of games to keep her entertained; however, there wasn't a single device that showed the time, and since she was underground, she couldn't tell the time by observing the sky either. During Follini's stay, her circadian rhythms were completely disturbed, leaving her life and daily schedules chaotic because she didn't know how long she slept or how long it has been since she last ate.

**Stefania Follini's living space -
Adapted from Steemit**

After 130 days under the cave, she finally gave up as the experience took a toll on her, physically and mentally. When asked to estimate how long she stayed in that cave, Follini said sixty days, which was off by a great margin.

**Stefania Follini coming out from the
cave - Adapted from Boing Boing**

I've tried to conduct the same experiment at home, but I only lasted eight hours before giving up, and when I guessed how long the experiment was, I was two hours off. This experiment doesn't prove that time doesn't exist; however, it shows how easy it is for humans to lose track of time if we were left in a world without clocks. It wouldn't be easy for you to get lost in your neighborhood because you are so familiar with it, but if you were left in the middle of dark and empty space, you wouldn't be able to tell where north is. This is the disorientation of space. Now visualize the world around you one minute ago and the world around you now. You wouldn't be able to tell which is which unless you look at your watch. This is the disorientation of time. These two examples further prove that humans rely too much on timekeeping devices, such as clocks and watches, so it may seem that time only exists inside clocks and are kept that way. Let's have a look at another possibility: A lot of people believe that the past and the future don't exist. Let me explain: At first, you may think, of course, the past exists. The past had already happened, and we cannot change what already happened. Are you sure, because all you know about the past is based on your memories? And we know that memories are subjective, so they aren't always reliable. Memories can be easily altered thanks to the Mandala effect, or worse, erased after an accident or just implanted not long ago. You may say I have proof of the past. I have photos, letters, and certificates. However, those are all subjective observations made by your mind in the present. How can you prove that the letters were written two years ago? How can you prove that the pyramids in Egypt were built by Ancient Egyptians instead of Egyptian time travelers from the future? I mean, sure, the pyramids look old, but aren't stage properties the same? This idea has many flaws, and I would hate to think that my happy memories and not-so-pleasant memories aren't real. The future seems just as real as the past, right? I mean we spend most of our time thinking and preparing for it. Students study so they can get a good job when they grow up, and adults work to get salaries. Despite our great expectations for the future, it seems that we are always "chasing it," but we never come close; the future is always one step ahead of us. This means the future always shows up at the right time, never early or late, but remember, the future is yet to happen, it is yet to be written, and it is yet to exist. This is the present theory; this suggests that the past and future are subjective and do not objectively exist. I don't support this theory;

however, I still decided to include it in this chapter for you to decide.

This is the end of this chapter. You may have noticed that I spent more time explaining that time is an illusion. I am not trying to prove that time doesn't exist, I am simply trying to say that time may not objectively exist, but I believe the concept of time in our subjective minds is still really beautiful and special. The question isn't and never was questioning the existence of time. Time doesn't exist objectively, sure, but why should that mean it's not real?

WHAT IS THE DIRECTION OF TIME?

I've mentioned in the previous paragraphs that time always travel in one direction, forward, and I realized I have to be more specific here: Time only travels in one direction subjectively, but it does not have to be the same in the objective world. After all, humans haven't developed the skill to completely push aside the influences of our surroundings yet; therefore, we neither see the true and full shape of our universe and how it operates. The world we live in could indeed be like the world in *Tenet* (a science-fiction action film), in which the direction of time can indeed be reversed. You may think that if time is traveling backward, of course, you would know. However, don't be so sure. If I were to show you a reversed footage of cracking an egg, you would definitely notice that the video has been reversed, but if I am to show you a reversed footage of a pendulum swinging from left to right, I don't believe you would ever notice anything wrong with the video.

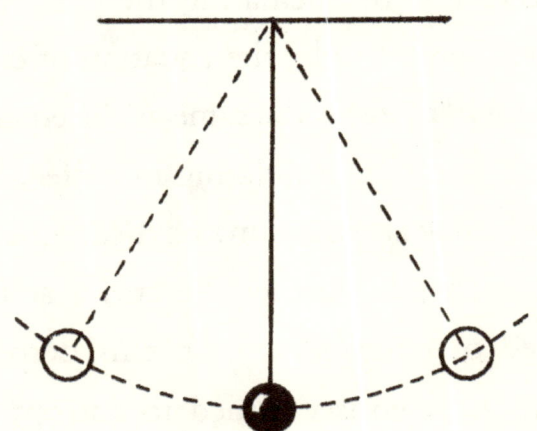

**The concept of a pendulum -
Adapted from Medium**

Scientists, cosmologists, and philosophers have been trying to answer the same question, which direction does the arrow of time point to, for centuries, but they never found a way to prove that time travels in this direction or the other. That is because even the most basic and fundamental laws of physics don't involve the direction of time. The laws relate to what happened previously and what will happen next. Therefore, we don't know whether we are calculating the past or the future. For example, the equation of cream melting in coffee is the same as the equation for the cream consolidating in coffee. So if we only look at the cream's physical equation, we cannot tell whether the cream actually melted or consolidated. In this desperate situation, scientists turned to entropy, the only thing in this world that qualifies to represent time. Entropy didn't let us down, and we soon realized that since the birth of the universe, the level of entropy has always been increasing in the "natural" world. By natural world, I mean the lifeless world that consists of rocks, water, metals, and so on.

However, in the "lively" world, the world that consists of plants, animals, insects, the level of entropy is less than that of the natural world. For example, nature turns cities into ruins over time, and life forms, such as humans, turn ruins (high entropy) into cities (low entropy). So the direction of time in the two worlds are different; the flow of time in the natural world is "backward," relative to the flow of time in the human world, the lively world. In the natural world, there may be an effect before a cause, opposing the lively world, where the cause causes an effect. Furthermore, like I mentioned in the last chapter, time is a feeling, and this feeling was created in the first place because our minds noticed something propelling the occurrence of events in this world, and this sense isn't always so accurate. Maybe, just maybe, from this moment on, time in our world is starting to spool backward, but we cannot even notice it.

Everyday folks like us don't need to bother too much about the change in the direction of time since it barely affects us. However, it's rather a big problem for physicists since they are the ones who came up with the laws and equations. I would like to end this chapter with a metaphor: Time is a river, and we are the rocks; the flow of the river may change, but we stay still.

THE SPEED OF TIME

Even though there are clocks and watches we use to keep track of time, it may sometimes seem that time passes at different speeds on different occasions. There are two reasons why this happens, and they explain why the speed of time doesn't stay constant and sometimes "differ."

The first reason for this is something I believe we are all familiar with. I would like to call it the misperception of time. Like I mentioned in the previous chapters, we humans aren't very good timekeepers, and our circadian rhythms can be easily disturbed with the absence of something we call clocks. Have you thought why time *seems* to pass at a speed slower than the slowest snail when you're working or doing the plank, and why time can pass faster than the blink of an eye when you are having fun and enjoying yourself at a party? Occasions like this happen quite frequently, and it feels like our daily lives are auto shows filled with cars ranging from 10 to 200 kilometers/hour. Imagine yourself as a child sitting in a classroom, half-listening to your teacher talking about a difficult math question. You glance at the clock now and then, but the hands on the clock seem to have fallen asleep. In this case, the passage of time you experienced is slower than a turtle and sloth combined. Now imagine you've just arrived at home after a long school day and decided to watch your favorite show; each episode is twenty minutes, but after watching one episode, it only felt like a short ten minutes. By looking at the two examples from above, we come to the conclusion that time seems to move faster when you're doing something enjoyable and slower when you're bored. This phenomenon gives rise to a popular quote that adults and teachers often say to children. You've probably already guessed it. Yep, the quote is "Time flies when you're having fun." But why is this the case? Psychologists who have studied this phenomenon have confirmed that people do indeed perceive time differently depending on what they are doing and what they feel toward that event. If you are doing something boring and you are feeling reluctant toward it, such as cleaning the toilet, your mind doesn't really focus on the task; instead, it wanders off. You could be wondering what to eat for lunch, which hat you should buy as a present for Amelia. When you think about what exciting thing to do after finishing this unenjoyable task, you develop an urge, an urge to hurry up and start doing something different and better. Sometimes you can make an event come to a close by speeding up, but when you can't, for example, during a literature lesson from ten thirty to twelve thirty, you can only wish time to pass quicker by not taking your eyes off the clock. Studies have also shown that the more frequently you look at a clock, the slower time seems to pass. On the other hand, if you're doing something fun and enjoyable, something you are passionate about, such as gaming, you tend to be more engaged in it at the current moment. When you develop that devotion, your mind is fully focused on the happiness and satisfaction in the present moment, and it doesn't have time to think about the time. In a way, you "lose track" of time, and you will only realize how long you've spent building that sandcastle until you've finished building that sandcastle. However, this is only a part of the reason why time seems to pass at different speeds. In the next paragraph, we'll look at the question from a more scientific standpoint.

To truly understand the speed of time, we have to turn to Einstein's theory of relativity, and in this paragraph, I will do my best to explain one of the biggest scientific discoveries of all time. Don't drift off. I promise the theory of relativity has a lot to do with the speed of time, even though it seems boring and confusing. The theory of relativity consists of many parts, and we'll go through them one by one. First, we have classical relativity, and it states that nothing in this universe is at absolute motion or absolute rest. For example, imagine a train (we'll come across a lot of examples surrounding trains in this paragraph) with two children on board playing catch with a ball. We know that the train travels at a constant speed of 200 kilometers/hour, and since the children are moving with the train, to them, they'll feel like they're standing still. Now to the ball the two children are throwing, the ball moves at a constant speed of 10 kilometers/hour; however, in the views of their rich grandmother who is standing still on the platform, the ball is traveling at a speed of 210 kilometers/hour, which is the speed of the train relative to the grandmother plus the speed of the ball relative to the two children. However, in this case, things aren't so simple. We have to take into account that the grandmother can't stand still while observing the train because the Earth is orbiting around the sun and the sun is orbiting itself and the Milky Way galaxy, in which we live in, is moving, and this universe is also constantly moving. Nothing is absolutely at rest or moving; they are just relative to one another. Second, we have special relativity, time dilation. We have to accept the fact that the speed of light is the same for all observers, you'll see why. Imagine a beam of light bouncing between two mirrors, replicate this setting, and install one of it on the same train from the last example. However, this time the train will be traveling at a speed close to the speed of light. In the views of the passengers on board the train (including the two children from the last example), the setting looks the same on board or off board the train. However, in the eyes of the two children's rich grandmother on the platform, the beam of light would be moving in a triangular trajectory, which means the distance it travels, if further compared to the beam of light from the setting on the ground next to the grandmother. Here comes the mind-boggling part: Both beams of light bounce off the mirrors at the same time. The only way for this to work is that the time must have *sslloowweedd* down on the ground because of the high speed on the train, so we come to the conclusion that the faster you move, the slower your surroundings move, and also, the slower you move, the faster your surroundings move.

Third, we have another special relativity, length contraction, and it is similar to time dilation. You see, time doesn't dilate enough to account for the speed of the light beams (from the last example) to remain the same. When an object moves at a speed close to the speed of light, not only does time slow down, but the object itself contracts as well. In our case, on the upgraded train that can almost reach the speed of light, the distance between the two mirrors contracted, and therefore the two beams of light, didn't have to travel as far.

The twins' observation of the setting

Grandmother's observation of the setting

Fourth, we have still another special relativity, the speed of light. In his famous 1905 papers on special and general relativity, Einstein stated that light is the fastest thing in the universe, nothing can travel faster than the speed of light, and the speed of light is always the same, relative to anything. Let me explain, imagine two trains traveling side by side at the speed of light. Child A on board Train A should see Child B on board Train B standing still because they are traveling at the same speed. However, according to Einstein, Train A is traveling at light speed relative to Train B, and Train B is also traveling at light speed relative to Train A, and they are both one light speed ahead of the other, so Train A is faster than Train B, and at the same time, Train B is faster than Train A. I know this is hard to digest, but it's an objective fact, proven by countless experiments. In one of the experiments, scientists measured the time for two light beams to reach Earth, and the light beams were sent out from two locations, one closer to the Earth than the other. Theoretically, the light beam closer to the Earth would reach Earth first; however, the result was that both light beams reached Earth at the same time. In conclusion to special relativity, time dilation and length contraction work together to make sure that the speed of light is the same, relative to anything or anyone. If the speed of light is relative then the speed of time must be relative as well. Fifth, we have general relativity, which states that the universe is a four-dimensional fabric formed by space and time, and Einstein calls this four-dimensional net space-time. Space can be affected and curved by an object that has a great mass, for example, the sun. Think of it this way: The space-time net is a trampoline, and the sun is a heavy rolling ball, and when you put the rolling ball on the trampoline, it makes a dip, a curve. Einstein's general relativity theory not only shows that space-time can be curved, but it also completes Newton's law of universal gravitation. According to Einstein, gravity is not a mysterious force attracting objects to one another. Let me explain: Let a ping-pong ball represent Earth, and what happens when you put it on the trampoline (space-time fabric) with the rolling ball (the sun) in the middle? The ping-pong ball moves toward the rolling ball, right? The greater the mass of an object is, the greater the curve it creates on the space-time net, and the curve in space can be experienced as gravity.

**The concept of spacetime -
Adapted from Medium**

On the other hand, the curve in time changes the speed of how quickly time flows. For example, in 1971, a group of scientists was trying to conduct an experiment to prove that Einstein's theory of relativity is correct. The experiment is called the Hafele-Keating experiment, named after the physicist and astronomer who conducted the test, Joseph C. Hafele and Richard E. Keating. This experiment includes three sets of four highly-accurate cesium-beam atomic clocks that were synchronized on the ground before the experiment started. The first set of clocks was taken on a plane and flown around the globe eastward, the second set of clocks was taken on another plane and flown around the globe westward, and the third set of clocks remained at the United States Naval Observatory. When reunited, the first two sets of clocks flown around the globe were found to be relatively faster compared to the third set on the ground, and their differences were consistent with the predictions of special and general relativity.

The Hafele-Keating Experiment - Adapted from Wikipedia

Albert Einstein - Adapted from Wikipedia

The more massive an object is, the slower time passes around it. Time on Earth is relatively faster compared to time on Saturn because Saturn has a greater mass than Earth, and at the same time, time on Earth is relatively slower compared to time on Mars because Mars has a smaller mass than Earth. So the scene in the 2014 movie *Interstellar*, in which astronauts found that seven years have passed on their spacecraft when they've only stayed for an hour on a planet not far away from a black hole (something of great mass), is actually based on scientific facts. So maybe, just maybe, the big bang could have only happened seconds ago if you observe it near a massive black hole at the center of the universe, but it's been 13.8 billion years for the people on Earth.

The first photo of a black hole captured by humans - Adapted from The Conversation

We come to the conclusion that there is no universal measurement of time; one second on the sun could be one year on Earth. Our sun in the solar system is relatively small compared to other suns in other planetary systems, so the time in our universe is relatively faster than other planetary systems, and that could explain why there aren't traces of higher civilizations because they haven't had enough time to evolve into beings such as humans. On the other hand, we may find civilizations superior to ourselves in the quantum world, where everything has less mass and time passes relatively faster. And remember, time dilation and length contraction can be experienced because of the effects of gravity and velocity.

So this is the end of this chapter, and note that above is an oversimplified explanation of Einstein's theory of relativity, but you get the point. According to Einstein, time travel is possible, and that will be our topic for the next chapter. Fun fact: Your head is slightly, just slightly, older than your feet because of the Earth's gravitational pull.

TIME TRAVEL

Whoa, time travel, big subject. Before we start discussing whether time travel is possible and how to time travel, we have to know why we want to travel through time so badly. We humans have come so far, and we have mastered the skills to "control" fire, water, earth, iron, electricity, and a lot more. We've always wanted to master time, the most beautiful and astonishing element that makes up our universe. Even as individuals, we have never stopped dreaming about changing an event in the past or viewing the wonders of the future. Time travel is humanity's great aspiration throughout time.

For time travel to exist, time must be an objective existence, which opposes what most philosophers believe. I'll talk about how time travel works and the different types of time travel plots in movies, shows, and books. Back to the topic, is time travel possible? The answer is yes, theoretically, according to Einstein's theory of relativity. Time travel is indeed one of the answers to the relativity equation. However, one of the greatest minds of this century, Stephen Hawking, believed that if a time machine is built, the whole world would come to an end. Why? Let me explain through a simple example: The big bang kicks off the existence of our universe, and I believe anyone with a time machine would want to travel to seconds after the universe was created. That creates a problem: You'd have countless future tourists from both the distant and not-so-distant future all packed in a relatively-small space, and that would cause the universe to explode. To prove his theory, Hawking hosted a party in 2009 for time travelers from the future, and he only sent out the invitations after the party finished so no one could have joined the party unless he/she traveled through time. And no one came. So there are either super strict time travel laws and time cops or humanity went into extinction before managing to build a time machine, or worse, the reason for humanity's extinction is because of the time machine. On the other hand, there's a lot of proof that time travel does exist or will exist, but that's for the next paragraph.

Hawking's party for the Time Travelers - Adapted from Atlas Obscura

Travel to the Past - Wormholes

Time travel has been and still is a very popular subject and a recognized concept among philosophers and physicists. At times countless books, movies, comics, theories, stories, TV shows, scientific papers surrounding time travel have been introduced to us, and they all have something in common: To time-travel, you'll need a time machine. The idea of a time machine was first popularized by H. G. Wells's 1895 novel, *The Time Machine*. But that's not the topic; there are two main types of time travel: the first one being what I call real time travel, where your actions as a time traveler can actually affect your past, present, and future; the second type is what I call fake time travel, where you can't change your past, present, and future. I will use plots from famous time travel stories to represent the different types of time travel.

Real time travel is such an astonishing subject that I get overexcited every time I talk about it. RTT can indeed be achieved if we follow Einstein's relativity equations, but the theory of relativity is much more difficult to understand compared to an Ikea instruction manual. So in the next few paragraphs, I will try to break it down for you, and you can refer to the contents list below:

Real Time Travel

1. Travel to the Future - Time Dilation
2. Travel to the Future - Cryopreservation
3. Travel to the Past - Wormholes
4. Travel to the Past and Future - Movie plots

Travel to the Future - Time Dilation

It would be relatively easier to travel to the future compared to traveling to the past. If you can reach the speed of light, you would be able to jump forward in time. What this means is that your time becomes relatively slower compared to others moving at "regular speed," and their time is relatively faster compared to yours. When you travel at such speed, a year to you could have been five years to an observer, who is traveling at a relatively slower speed. And in this case, you actually traveled four years into the future because 5 minus 1 is 4. Despite how easy and convenient this method may seem, the fastest man-made transport cannot even reach 1 percent of the speed of light, so traveling to the future through the concept of time dilation remains a dream.

Travel to the Future - Cryopreservation

Cryopreservation is probably the most simple way to travel into the future. It preserves your body through freezing, and your metabolism will be slowed down, so in a way, you will not age. The process is like a deep sleep; when you wake up again, you'd find that your surroundings have changed, your family members have aged, years have passed since you were frozen, but the good thing is you wouldn't be much different compared to yourself when you were frozen. However, this technology has only been used on dead or dying people because some scientists and biologists suggested that dying is a process and freezing the "dyer" will slow that process down. So the idea is for the "dyer," who has severe incurable diseases, to be unfrozen in the future when medical care and technology is more advanced. Since this technology is still in its early stages, freezing a living person is illegal because cryopreservation can have negative effects on a human's mind and body.

Travel to the Past - Wormholes

I have already introduced two scientific ways to really time travel to the future, and I know you are probably more interested in traveling to the past. Jumping into a wormhole should technically allow an object to travel to the past or future, and it is permitted by Einstein's theory of relativity. Black holes are "holes" in our universe with great mass and energy. They further increase their mass by sucking objects in with their high gravitational pull. Not even light can escape from black holes. White holes are portals in our universe with great mass and energy, however different from black holes. White holes emit objects, and nothing can enter it. Black holes and wormholes are similar, except for one thing: A black hole leads to a dead end, but a wormhole creates a portal between two points in space-time, one of the points being a black hole and the other being a white hole. Objects go in from the black hole and go out from the white hole, this is the Einstein-Rosen bridge (a fancy name for a wormhole). A wormhole suitable in mass should allow a living human to travel through both space and time. A wormhole can be visualized as a tunnel with two ends at separate points in space-time.

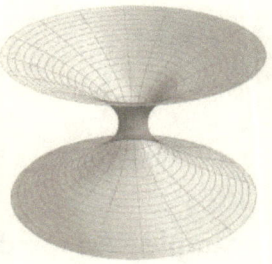

Concept of Einstein-Rosen Bridge - Adapted from Initiative for Interstellar Studies

A wormhole can be created using a certain special configuration of matter and energy, which should create an enormous "dip" in the space-time fabric and allow the formation of a tunnel. Think of it this way: You have a piece of paper with two dots drawn on opposite sides. The shortest distance between the two points would be for you to fold the paper in the middle and poke a pencil through one of the dots into the other, the length of the pencil is the shortest distance.

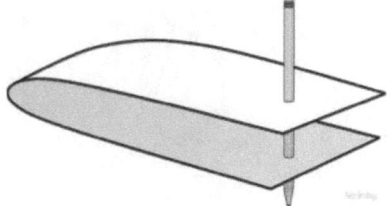

Paper example - Adapted from Bard Course Blogs

For now, wormholes only exist hypothetically inside our minds, and it would be hard to create and harder to sustain them because gravity tries to squeeze it shut as soon as it gets created. To keep a wormhole open, we need something called exotic matter, and exotic matter is even more illiterate compared to a wormhole because it hasn't been found existing naturally in our universe yet. Exotic matter is different from matter and antimatter because matter and antimatter are attractive, they cluster together; but exotic matter is negative and repellant in nature (similar to how two north poles on magnets can never stick together), so exotic matter could be used to keep a wormhole open. However, even if we manage to create a wormhole, we still can't travel to the Jurassic period and ride dinosaurs or travel to the year 2077 and drive flying cars, partly because of the gravitational force that can break our body when we enter a wormhole that is not big enough in mass, but mainly because we can only travel to the past after that wormhole was initially created, and we cannot travel to the future. For example, if a wormhole were created in 2021, we could be expecting time travelers from the year 2022, but ourselves cannot travel to the year 2022 nor the year 2020 because that was before the creation of the wormhole. Wormholes are like shortcuts in space-time, it's like a hack, and they allow us to travel faster than the speed of light.

Time Travel to the Past and Future - Movie plots

This paragraph is nothing like the other three; it consists of more unscientific and abstract ways to travel through time. (And they are all based on plots adapted from fictional productions.) First, we have the movie *Justice League: Snyder Cut*, where you can reverse events to their original state by traveling at a really-fast speed. Second, we have the book *The Time Machine*, where you can build a time machine (somehow) and use it to travel through time. Third, we have the movie *X-Men: Days of the Future Past*, where your present mind can travel into your past and future body. Fourth, we have the movie *The Time Traveler's Wife*, where you are gifted with the superpower to travel to the past and the future with your present mind and body. Fifth, we have the book and movie *Harry*

Potter and the Prisoner of Azkaban, where you completely rely on a magical device to help you travel through time. I know these movie plots may seem really absurd, but can you prove they are not real?

In a sense, RTT may seem better than FTT; however, RTT creates paradoxes that are hard to solve and explain. The most well-known paradox is probably the grandfather paradox; it tells the story of a grandson (let's call him Bob) who was extremely unhappy with his family and himself. Thus, Bob created a time machine and traveled to the past before his father was born, where Bob found his younger grandfather and killed him. Therefore, Bob's grandfather was dead and Bob's father never existed and Bob never existed. However, if Bob never existed, then who killed his grandfather?

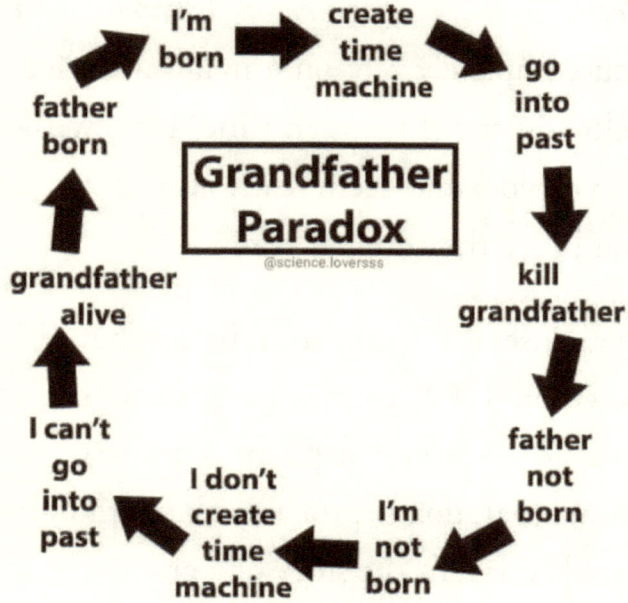

The loop created by the Grandfather Paradox - Adapted from Science loversss

There are solutions to these paradoxes, one of them being Novikov self-consistency principle, and it states that if an event will create a paradox, then the event has no chance of happening. Ummm . . . that doesn't seem so convincing. There is another similar solution, and that is the banana peel mechanism, and we can apply it to the grandfather paradox. Same story, same grandson, same old Bob. Bob carried a gatling pointed at his younger grandfather, but just when he was about to shoot, he stepped on a banana peel, slipped over, hit his head against a sharp rock, and died. Grandfather lived, and Bob and his father were born.

A banana peel - Adapted from ABC Science

The banana peel mechanism states that something will always stand in your way from changing the past and disturbing causality. However, the banana peel mechanism is just unilateral and subjective because you will always make changes to the past through RTT despite how "big" those changes are. There is one more solution, and that is the hypothesis that time is a loop. In this loop, cause and effect simply chase after each other. This concept can be shown in a simple example: Imagine that you receive an instruction manual of how to build a time machine. You decide to give it a try, and you took twenty years to construct it. Suddenly, you understood something and jumped on the time machine and carried the instruction manual with you. You traveled to the day you received the instruction manual twenty years before and left it on the doorsteps of your younger self. So you built the time machine because you received the instruction manual, and you received the instruction manual because you built a time machine. This is the bootstrap paradox, where an object (the instruction manual in this case) or person or information has no origin in time. In this case, time doesn't look like a line but a circle instead, and since time is a loop, attempting to change the past and future is pointless since the fates of all things are set. This also explains the question of whether the chicken existed first or the egg existed first. The chicken existed because the egg existed, and the egg existed because the chicken existed. I am actually a big fan of this time loop theory, and I came up with my own theory surrounding the existence of our universe. Many scientists believe that at one point in history, our universe will eventually "die" and collapse into a dimensionless singularity, back into how the universe started before the big bang. Here's how my theory goes: Our universe was created because our universe "died," and our universe "died" because our universe was created. The end of our universe in the future is the starting point of our universe in the past, and sadly, if my theory is correct, then all our fates are set.

Now we'll finally be discussing fake time travel. Don't judge it by its name. It's still time travel in a way that doesn't allow you to change the events in history. General and special relativity are laws surrounding the super big world, such as black holes and planets; quantum mechanics are laws surrounding the super small world, such as atoms and electrons. If there is a way for a person to shrink to a size smaller than or close to the Planck Length, the fundamental laws of physics wouldn't apply to you, and with supportive technology, you could travel in time. However, you wouldn't be traveling through the timeline in your world. For example, if you go back to the year 2019 and warn everyone about Covid-19, you will create a separate universe and a separate timeline where the pandemic never happened, and this is the many-worlds theory. The reason you wouldn't be able to change your past is if you travel to the past, that past becomes your new future, and your former present becomes your past, which can't now be changed by your new future. Time travel may also be achieved by traveling in your memories or in a giant database that stimulates and predicts the events of the world.

Many Worlds Theory - Adapted from Project Q

In conclusion, time travel isn't a technology humans are capable of grasping yet, and we have to understand it before applying it to real life. If time travel is achievable, my only wish is to be an observer who can see all the details of the past.

CONSPIRACY THEORIES SURROUNDING TIME TRAVEL

Since this is the last chapter of this book, I thought, why not choose a topic that is more fun and thought-provoking for all of us? In this chapter, I will be discussing the "conspiracy theories" surrounding time travel. No, this has nothing to do with the flat Earth theory. I found all the theories online, and I apologize if I used your source without permission.

1. *The Moberly-Jourdain Incident* - The story started in 1901, when two middle-aged English academics, Charlotte Anne Mobley and Eleanor Jourdain, visited the Petit Trianon, a small château in the grounds of the Palace of Versailles. Mobley and Jourdain claimed in their

later published book that they ended up 120 years in the past and that they hung out with Marie Antoinette. (I guess they had cake and tea together while discussing smelly underclass farmers.) Mobley and Jourdain's book about their unbelievable experience, *An Adventure*, became a best seller when it was first published in 1913. This story is quite absurd, but who knows and who cares whether it's real or not.

Photos of Mobely and Jourdain - Adapted from Wikipedia

The Adventure - Adapted from AbeBooks

2. *The Scottish Time Slip* - This story is somewhat similar to the first one, but instead of traveling back in time, it is about traveling forward in time. Victor Goddard flew into the future for a short period. Goddard flew planes for the RAF in World War 1 and World War 2, and he was the director of intelligence at the Air Ministry, so a

reliable source. According to Goddard, in 1935, he briefly flew into the future. It happened when Goddard left Drem Airfield near Edinburgh to fly home, then he came across some pretty bad weather and decided to fly back to Drem; however, as he approached the airfield, it looked completely different. There were men wearing blue uniforms and planes painted yellow, a sight Goddard never saw before.

Victor Goddard with his plane - Adapted from Fandom

Goddard flew away without landing, and four years later, the air force changed all their uniforms to blue and painted all their planes yellow. Curious, very curious indeed.

3. *John Titor* - One of the most famous, if not the most famous, time traveler in history. It all started when an account named John Titor suddenly appeared on American social media and claimed that he is a soldier from the year 2036. At first, no one believed him because nobody is that stupid. But when Titor posted photos of his "time machine" and explained how it works, several academics claimed that Titor's "time machine" is really advanced and that he could indeed be a time traveler.

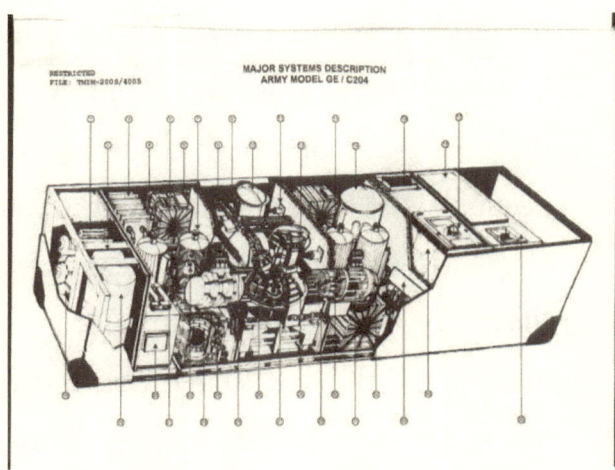

**John Titor's time machine design -
Adapted from Digitized Graffiti**

After hearing the academics' remarks, more and more spotlights were shone on Titor, and he became the hottest topic that everyone was talking about nationwide. Once Titor became famous, people asked him a lot of questions about himself and about the future. Titor did not disappoint us and answered most of them flawlessly, and he seemed too smart to be a swindler as he had a good understanding of time travel, physics, and science. Titor claimed that he is on a mission to retrieve an IBM 5100 computer, which he said was needed to debug various legacy computer programs in 2036, a possible reference to the UNIX year 2038 problem. According to Science ABC, the 2038 problem refers to the time-encoding error that will occur in the year 2038 in 32-bit systems. This may cause havoc in machines and services that use time to encode instructions and licenses. The effects will primarily be seen in devices that are not connected to the Internet. Why was the IBM 5100 computer so important? Because the IBM 5100 runs APL and BASIC programming languages, and it ceases to exist in 2036. Titor claimed that his grandfather participated in the design and creation of the computer.

The IBM 5100 - Adapted from Wikipedia

We later learned that Titor didn't travel to 2000. He traveled to 1998, when he lived with his parents and his younger self secretly together. Titor also claimed that the time machine he used was created by CERN in 2034. CERN, the European Organization for Nuclear Research, is one of the world's largest and most respected centers for scientific research. CERN announced that it would start constructing a giant particle accelerator that had a great chance of creating an artificial wormhole. Titor also said that his time machine works by curving space-time through creating an artificial singularity (the center of a black hole). It all makes perfect sense, right?

CERN's particle accelerator - Adapted from Universe Today

Titor didn't stay for long; he disappeared on the Internet after two years, leaving mysteries unsolved until now. Let's look at Titor's "prediction" about the future and, while you're reading this, the past. Titor successfully predicted the 2001 earthquake in Peru, which happened five months after his prediction was made. Titor also successfully predicted that China would soon achieve manned flight, and it really did happen on October 15, 2003. Titor also predicted that half of Japan would become a redistricted area by 2020 because of nuclear contamination and the new capital would be Okayama instead of Tokyo; and even though his prediction didn't come true, however, in a way, he did predict the Fukushima Daiichi nuclear disaster a decade before it happened. There are reasons why Titor's predictions didn't come true; some say it's because of his arrival, and Titor also claimed himself that through time, the universe he came from and our universe would become more and more different. Our time traveler also said that WW3 broke out in 2015 in his universe, and we all know it didn't happen, phew. Since John Titor's tales attracted so much attention, people soon noticed a book called *John Titor: A Time Traveler's Tale* published by a lawyer in Florida named Larry Haber.

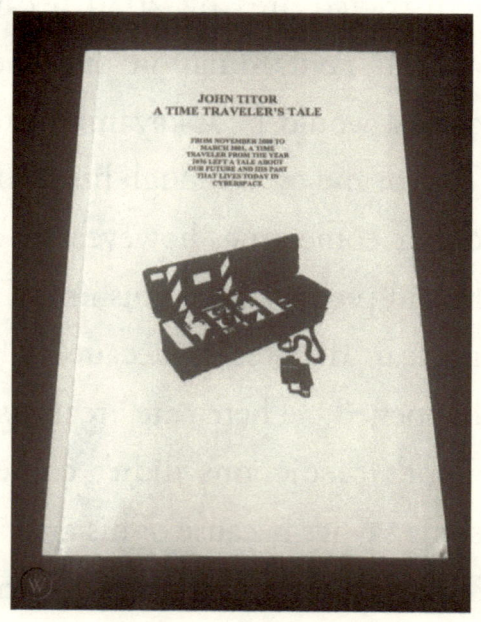

John Titor A Time Traveler's Tale - Adapted from WorthPoint

Larry Haber - Adapted from People-Background-Check

Haber claimed that a woman contacted him in 2003, and guess who that woman was, yep, John Titor's mother, and Haber wrote this book based on the information she provided. Many people started to doubt whether John Titor is real at all or if he is just a fictional character created to help Harber sell his books. Haber's brother Morey is a computer scientist, and there is a possibility that this fabricated story was created by their hands. There are rumors that the IP address of John Titor's account is the same as the IP address of the for-profit John Titor Foundation, which Haber is a part of. Hmmmmm . . .

4. *Andrew Carson* - In 2003, according to *Weekly World News*, a man named Andrew Carson was arrested by the FBI, and he was accused of Securities and Exchange Commission violations because he turned an initial investment of $800 into $350 million through investing in stocks at a time when the stock market was unstable. The FBI couldn't trace Carson's origins before 2002. The police interrogated Carson, and he claimed that he is a time traveler from the year 2256 and prerecorded the data of the stock market in 2003 to

profit from it. Just like his other time traveler friends, Carson also "kindly" provided us with information of what would happen in the future. Carson claimed that the Iraq war would start on March 20, 2003, and Bin Laden would be hiding in a city in Northern Pakistan; both his claims are true. Impressive, right? Carson, like other time travelers, didn't stay for long and vanished in 2003 after a stranger paid a bail of $1 million to get him out of prison. *Weekly World News* first published an article about Andrew Carson, and we later found out that *Weekly World News* is a very unreliable source, known for publishing contents such as Donald Trump being an alien. Other more reliable news agencies only published reports about Carson weeks after the Weekly World News.

So Mr. Carson could have been one of *Weekly World News'* exclusive reports.

The Weekly World News' Article about Andrew Carson - Adapted from CR Berry

5. *Sergey Ponomarenko* - On April 23, 2006, a young man dressed in clothes

from the last century appeared on the streets of Kiev, the capital of Ukraine. The young man carried an "ancient-looking" camera with him. He looked around, confused, and didn't know what to do and where to go. It didn't take long before the young man's suspicious appearance was noticed by the police officers on duty nearby. The young man was asked to show his ID card, his name is Sergey Ponomarenko, and he was born in 1932, and his ID card was issued by the Soviet Union.

Sergey Ponomarenko's ID card - Adapted from Quora

The police officers thought Ponomarenko had psychological issues, so they brought him to a psychiatric hospital, where the psychiatrist, Pavel Krutikov, asked Ponomarenko questions about his identity. Ponomarenko claimed that he was born on June 17, 1938, and that he was twenty-five years old. He also said that the last thing he remembered was it being midday on April 23, 1958. Ponomarenko seemed worried when he was told that he somehow stumbled into the future. When asked to recall how he ended up in our time, Ponomarenko said he was taking a walk during his break, he was outside, and he suddenly saw a strange object in the shape of a massive bell hovering in the air, and it was flying along a strange trajectory. He captured the sight with his camera, and before he knew it, he was in the future. A hobbyist photographer himself, Krutikov recognized Ponomarenko's camera being a Yashimaflex, the first model ever manufactured by the Japanese company decades ago; however, the camera looked brand new.

The Yashima flex - Adapted from Camera-wiki

Krutikov took the camera and rushed to develop the photos. He watched as Ponomarenko entered a clinic room to wait for him, and this scene was also recorded by the security camera. When Ponomarenko entered the waiting room, he and the receptionist noticed that the time on the clock hanging on the walls of the clinic was one hour late. Not a long time passed before Krutikov returned with the photos, only to find that Ponomarenko disappeared inside the room. Krutikov went to the Civil Affairs Bureau and found that Ponomarenko did indeed exist, but he disappeared in the 1960s. To further trace his mysterious patient, Krutikov used the photos to help him. There was a photo of Ponomarenko with a woman, possibly his girlfriend or wife.

Photo of Ponomarenko with the woman - Adapted from inf.news

The same woman from Ponomarenko's photo is 2006 - Adapted from inf.news

The photo of the mysterious bell shaped object Ponomarenko saw - Adapted from Ava's Garden

Krutikov found the woman, who was already seventy-four years old at the time, and she claimed that Sergey Ponomarenko did indeed exist, but she never saw him again after his disappearance in the 1960s. Sergey Ponomarenko is the only official time traveler, and his story sounds more real compared to the other four, but still, it could be a hoax after all.

Congratulations! You managed to reach the end of this book. My advice to you is to relish the past, live in the present, and always look forward to the future.

Farewell, until next time...

www.ingramcontent.com/pod-product-compliance
Lightning Source LLC
Chambersburg PA
CBHW030838180526
45163CB00004B/1371